DATE DUE

DATE DUE

DEMCO 128-8155

14875

595.1   Halton, Cheryl M.
HAL   (Cheryl Mays)

Those amazing
leeches.

THOSE AMAZING
LEECHES

THOSE AMAZING
LEECHES

By Cheryl M. Halton

DILLON PRESS, INC.
Minneapolis, Minnesota 55415

To John for all his encouragement and support,
and to Charles and Andrew

Photographic Acknowledgments

The photographs are reproduced through the courtesy of Timothy Branning (pages 6, 10, 33, 57, 71, 81); Carolina Biological Supply Co.; Department of the Army; Robert Gillette (pages 20, 65, 67, 69); John Halton (pages 16, 30, 76, 77, 91, 94, 98); National Library of Medicine; John Nuhn (page 87); Dr. B.W. Payton (cover, pages 26, 59); Smithsonian Institute of Medical History; and AP/Wide World Photos.

Library of Congress Cataloging-in-Publication Data

Halton, Cheryl Mays.
  Those Amazing Leeches : Cheryl Mays Halton.
   p. cm.
  Bibliography: p.
  Includes index.
  Summary: Explores the physiology, habitat, and past and present medical uses of a variety of leeches.
  ISBN 0-87518-408-1

  1. Leeches—Juvenile literature. [1. Leeches.] I. Title.
QL391.A6H27 1989
595.1'45—dc 19
                              88-35908
                                    CIP
                                     AC

Dillon Press, Inc., 242 Portland Avenue South
Minneapolis, Minnesota 55415

Printed in the United States of America
  2 3 4 5 6 7 8 9 10 98 97 96 95 94 93 92 91 90

About the Author

Cheryl Mays Halton is a freelance writer and the former Director of News and Information for the University of Texas Schools of Nursing. During her research for this book, she interviewed medical professionals and researchers and visited Dr. Roy Sawyer at the world-famous Biopharm in Swansea, Wales.

Ms. Halton earned a bachelor's degree in zoology and a master's degree in communications from the University of Texas at Austin. She resides with her husband and two children in Austin.

CONTENTS

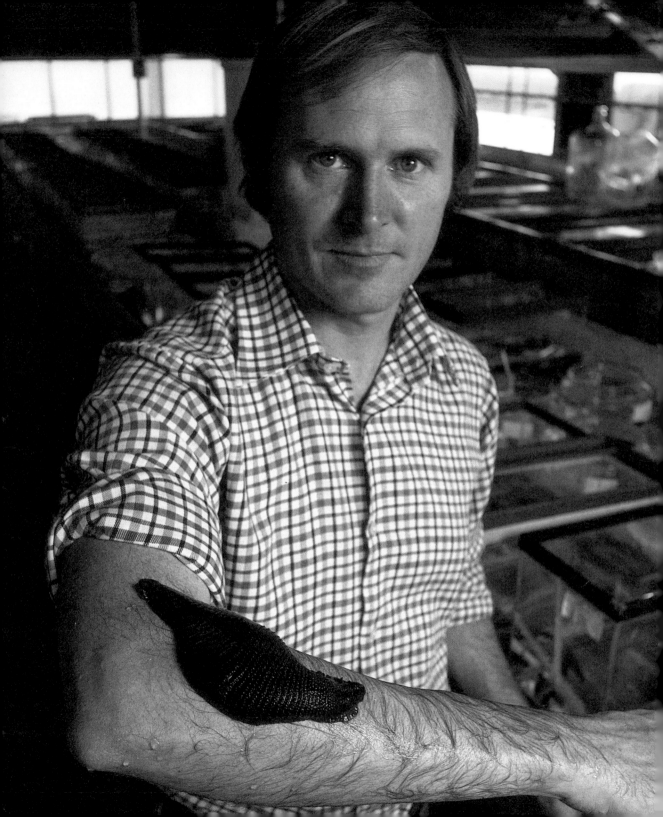

CHAPTER
~1~

Leeches as Pests

For six long weeks in 1977, he had braved scorpions, tarantulas, poisonous snakes, and the threat of tropical diseases. Now standing to his waist in the thick, slimy water of a steamy, tropical marsh in French Guiana, the zoologist was thrilled. He had found a giant species of leech that was thought to be **extinct.*** Most people would have screamed in terror as its 13 inches (33 centimeters) slithered by. Instead, Dr. Roy Sawyer yelled with delight as he reached out to capture it.

In the minds of many, leeches are creepy, crawly creatures, good only for adding a touch of realism to low-budget horror movies. Yet these little-known animals, popular with doctors

*Words in **bold type** are explained in the glossary at the end of this book.

◀ *Dr. Roy Sawyer with a giant Amazon leech on his arm.*

8

long ago for the wrong reasons, are increasingly important to physicians and scientists today for the right ones.

Leeches Everywhere

Most leeches are bloodsucking **parasites**, which means they live off of other animals without helping the host in return. Others are **predators** that swallow their prey whole.

Leeches range in length from a quarter-inch (.6 centimeters) to 18 inches (45.7 centimeters) long, although most are 1 to 2 inches (2.5 to 5 centimeters) in length. Their muscular bodies can stretch to give them a thin, snakelike shape or draw up so compactly that they look almost round. During a meal lasting just a few minutes, they can easily double or triple in size as they consume up to ten times their weight in blood.

Leeches come in many colors. Some are a dull brown or black. Others are more showy with bright stripes and spots of orange, red, green, and yellow. Some have branching gills that look like many fingers reaching out.

Most people think of leeches as tropical creatures that thrive in African jungles or swamps, or along the banks of slow-moving rivers such as the Amazon in the hottest part of South America. Leeches do live in those places. But they also are found in small desert waterholes, on mountains at heights of 12,000 feet (3,660 meters), and in polar oceans. Odd as it may seem, more leeches live in antarctic waters than in all the tropics.

Most, but not all, leeches live in water. Some are **amphibious**, living mostly in water but crawling out onto land on warm, humid nights in search of food. Others are land leeches common in damp regions around the world, such as Vietnam, Sri Lanka, the East Indies, Japan, and parts of China, India, Australia, and South America. They hide on the ground under decaying leaves or on wet plants.

A small crocodile with a large leech attached to its back.
A leech this size could easily kill a small animal.

Areas crawling with leeches in the wet season appear to be free of them in dry weather. The leeches are still there, but they have burrowed into the damp subsoil or are hidden under rocks near stream beds. With the first rain, they appear in the thousands. In areas that always are moist, leeches are active year-round.

Age-old Pests

As many people will testify, leeches can be a real nuisance to those who like to be out-doors. If you like to cool off by taking a dip in a lake or pond, or by wading barefoot in a stream, beware! You may find that a leech is enjoying the water, too, and has latched onto you.

For the most part, these leech attachments are merely irritating. But in some parts of the world leeches have been, and still are, a big problem for both people and animals.

In 1799, soldiers serving under the French general Napoleon marched from Egypt across the Sinai Peninsula to Syria. In the scorching heat they drank water from every source they could find, including leech-infested ponds. As a result, tiny leeches attached to the insides of noses, mouths, and throats of many soldiers. Though small when swallowed, the leeches enlarged greatly once they gorged on blood.

12

Many soldiers died from blood loss. Others died from suffocation when the blood-swollen leeches blocked their air passages. One hundred years later, during World War I, British troops met the same fate in the Sinai when they, too, drank from leech-filled water holes.

During the 1800s, coffee planters in Ceylon (now called Sri Lanka) wore "leech gaiters." Made of a densely woven material, the gaiters were worn on the legs to keep leeches from attaching. Natives without gaiters smeared their bodies with oil, tobacco ash, or lemon juice to repel leeches. These measures did not always work. Historians report that during a rebellion in Ceylon in 1818, when troops marched on foot through the mountains, soldiers and especially natives were attacked so severely by land leeches that many of them died.

More recently, American soldiers who fought in Vietnam reported that leeches were so thick in some parts of the jungle that they would drop

During the Vietnam War, U.S. soldiers in the jungles were plagued by leeches.

from the trees like rain when the vegetation was disturbed. Soldiers learned to keep all parts of their bodies covered and to cuff tightly their shirt sleeves and pant legs to keep leeches out. Yet to their dismay, leeches could find even the smallest rip in their clothing and could slip in to feed on their bodies.

14

Today, mountain climbers and trekkers on expeditions in the Himalayas often hike great distances through the rugged terrain of Nepal. Their constant companions are the land leeches, which the people of Nepal call *zurkars*. These creatures thread themselves easily through the lace holes of tennis shoes or hiking boots. Because their bite is painless, hikers usually are not aware that they have been attacked—until they see blood stains on their clothes, feel a lump getting larger in their shoe, or sense something heavy on their ankles. Hikers have reported that leeches have sometimes gorged undetected until they reached the size of bananas!

More Than Just Pests

Though today leeches are widely regarded as disgusting pests, for thousands of years people used them to treat many ills. During the height of their popularity in the nineteenth century, one species of leech was so overused in medicine that it nearly became extinct in Europe. With the spread of modern medicine, leeches fell from favor.

New research, however, has proven that leeches are more than just pests. In the medical and scientific communities, leeches are regaining their lost popularity. Physicians once more use leeches, especially in plastic surgery. Yet some scientists believe the leeches' most important contributions are still to come. They are used in studies that may one day reveal how nerves are formed and how they work in higher animals, including humans. Scientists hope other studies will explain how injured or destroyed nerve cells **regenerate** or regrow. Researchers are also studying substances found in leech saliva, which they believe will have many important medical uses.

Dr. Roy Sawyer, one of the world's leading leech experts, believes that these substances will

become important weapons in the fight against diseases that cause heart attacks and strokes. Sawyer also predicts that they will be used to treat glaucoma, an eye disease that can lead to blindness, and connective tissue disorders such as rheumatoid arthritis.

People haven't been misjudging leeches for all these years. They truly can be pests—even deadly ones. Yet these new and important ways of using leeches may cause many humans to view these remarkable creatures with new-found respect.

◀ *Dr. Roy Sawyer holds a pair of Amazon leeches. Sawyer, one of the world's leading leech experts, believes that leeches will have important uses in medicine.*

CHAPTER

~2~

Fascinating, Mysterious Leeches

Leeches are fascinating animals, simple in design yet capable of complex behavior. They can walk like inchworms using both front and rear suckers, or stretch themselves into long, graceful ribbons and swim like miniature sea serpents waving slowly up and down. They sense motion, pain, light, and odor, hunt their prey, and reproduce. All of these activities are controlled by a very simple nervous system, spread the length of their body, which is not much more than a muscular tube.

Leech Basics

The leech belongs to a large group of worms called **annelids**. It is a close cousin of the earthworm and shares many common traits. Both are **hermaphrodites**, meaning they have both male and female reproductive organs. Both also have an organ called a **clitellum** which **secretes**, or forms, a **cocoon**, a type of covering, to protect their fertilized eggs. Even their fertilized eggs, called **embryos**, have many similarities as they develop into worms or leeches.

The earthworm uses **setae**, small, stiff bristles along each side of its body, to move and burrow. Most leeches, however, have **suckers** at each end of the body for moving on solid surfaces and for attaching securely to a host for feeding. They use many strong body muscles for swimming carefully through the water.

An earthworm's body is made up of a series of connected rings, called **annuli**. The earthworm grows by adding more and more of these rings to the tail end of its body. Each ring is separated inside the worm by thin body walls.

Leeches come in many different sizes and shapes. Whether they are large, like this Amazon leech, or small, all have a body made up of thirty-two segments.

The leech, on the other hand, whether young or old, has thirty-two segments. Each segment has from three to five annuli. Unlike the earthworm, the leech does not have internal walls separating its segments because it needs a large area to store the blood it takes in.

The leech has keen sense organs which it

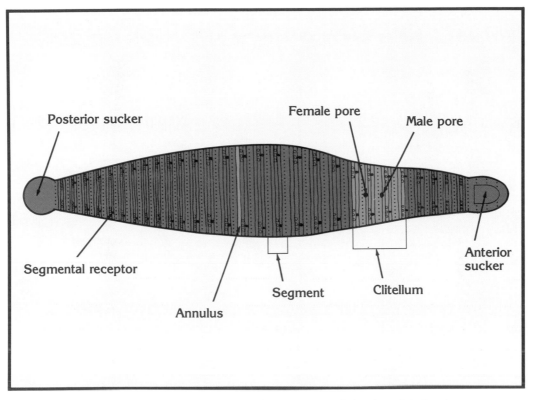

This drawing shows some of the parts of the leech's body.

uses to detect opportunities to feed. Very small colored dots or patches on the leech's body are sensitive to water or ground movements. Called **segmental receptors**, these organs are tuned to pick up even the slightest vibration. An animal or person entering a pond or walking along a jungle trail easily excites these sensors. Once

When a land leech detects a vibration, it stands erect with one end attached to the ground (left). *As the animal or person comes near, the leech's other end rises up, and the leech flips end over end until it reaches its prey* (right).

a land leech detects the vibration, it stands erect along the pathway with one end attached to the ground. The other rises into the air and lurches forward as the animal or person comes near. The leech then flips end over end until it reaches its prey. A leech that lives in water swims quickly toward the source of any water movement, hoping to attach to a tasty host.

A shadow cast by an animal, the warmth of its body, or even its breath will alert the leech. Individual light-sensitive cells, called eyespots or **ocelli**, are found on all parts of the leech. Eyes, which are groups of ocelli, are found only on the leech's head region. Both eyes and ocelli are stimulated by changing light conditions such as the shadow of an animal falling on the leech.

Other cells on the leech's body can detect even tiny amounts of substances such as skin oils, blood, or even carbon dioxide exhaled by an animal. When these sensory cells are stimulated, the leech moves quickly toward the target, ready to attach as soon as it reaches its prey.

How Leeches Feed

Bloodsucking leeches remove blood from their hosts using either a **proboscis** or jaws and teeth.

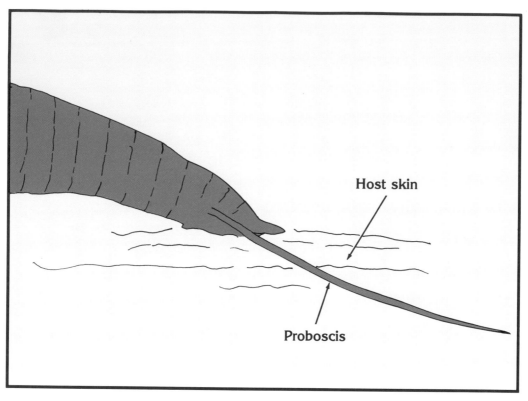

Host skin

Proboscis

This drawing shows how a leech uses a proboscis to draw out blood from its prey.

A proboscis is a sharp, needlelike organ that is inserted into the host to draw out blood quickly and efficiently. It can be up to one-third the length of the leech's body. In the case of the giant Amazon leech *(Haementeria ghilianii)*, which can grow to a length of 18 inches (45.7 centimeters), the proboscis can be up to

6 inches (15.2 centimeters) long.

Other leeches, including the medicinal leech *(Hirudo medicinalis)*, are armed instead with three jaws and as many as three hundred tiny teeth. Their bites produce shallow Y-shaped cuts on their host.

Both kinds of attachments, whether by puncture or by bite, sound painful, but surprisingly they are not. The saliva in all leeches (except the "stinging leeches" of Borneo and India) contains an **anesthetic** that numbs the wound. Because its bite is painless, the host is not alerted to the leech's presence, and it can feed undisturbed, at least for a while. The saliva also contains a variety of other chemicals. A "spreading factor" dilates, or temporarily widens, the host's blood vessels near the bite, thus increasing the flow of blood to the wound. Other chemicals keep the blood from **coagulating**, or clotting, before the leech has finished feeding.

The anticoagulant and spreading factor

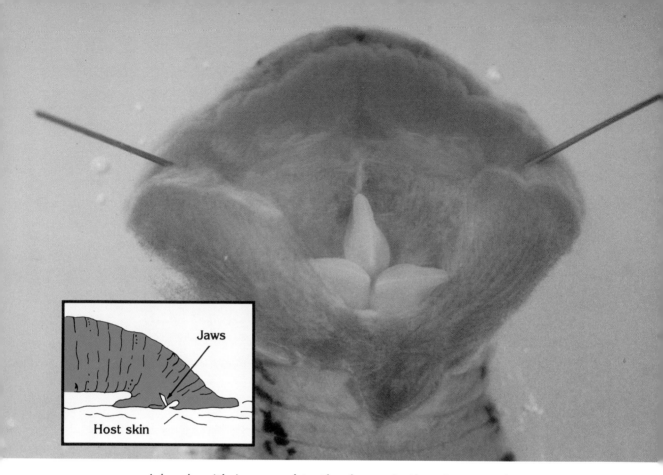

Jaws

Host skin

A leech with jaws and teeth, shown in the photograph above, feeds by biting its host and drawing out blood.

keep the wound bleeding long after a leech has dropped off. The anticoagulant also keeps the blood liquid once it is in the leech's digestive tract.

Leeches do not feed often. In fact, it is not uncommon for them to eat only once every six months, and they can go as long as a year without a meal. This is possible because when leeches do feed, they take in huge quantities of blood.

The medicinal leech can absorb up to five times its weight (about two teaspoons of blood) in about twenty minutes. Some land leeches can consume as much as ten times their weight in one sitting. Four or five large leeches can drain the life from a rabbit in half an hour.

As it takes in blood, the leech rids itself of the substances it doesn't need, mainly water and salt. A pool of this clear liquid can be seen forming around the land leech as it feeds. In the medicinal leech, almost half the weight of the blood is lost within ten days.

The leech retains only the red blood cells, digesting them over a long period of time. A special bacteria lives in the gut of the leech and attacks other types of bacteria that would spoil the blood before it could be digested. This bacteria also breaks down the blood cells one at a

Hirudo medicinalis, *the medicinal leech.*

time, which allows the leech to absorb the products of digestion very slowly. In fact, whole red blood cells have been found in the leech's gut as long as eighteen months after a meal.

Leeches are the only known animals that rely on bacteria, rather than chemicals called **enzymes**, to break down food. Even young leeches have this bacteria in their gut before they leave the cocoon. Scientists believe that the parent leech passes this bacteria into the cocoon as it is formed.

Forming Cocoons

Although leeches have both male and female reproductive organs, they rarely fertilize their own eggs. Rather, most form pairs for mating, and fertilize each other's eggs. Jawed leeches have a penis which they use to transfer sperm to another leech through its vagina, or female opening. Those with a proboscis usually lack a penis. When those without a penis mate, they wrap around each other, simultaneously attaching **spermatophores**, or sperm packets, to the body wall of the other leech.

Once fertilized, the eggs of both the jawed leeches and those with proboscises are ready to be placed in a cocoon. In some families of leeches, this takes place in just a few days. In

Cocoons of the medicinal leech.

others, it can occur as long as nine months later.

Some kinds of leeches lay their cocoons in water, secreting a sticky substance from their front sucker onto a protected surface such as a rock or log. Onto this sticky substance, the leech begins to secrete the cocoon from its clitellum. The outer wall of the cocoon forms as a narrow band that encircles the leech's body. The leech slowly rolls its body around and around within this bandlike cocoon to smooth the inside wall.

Once the wall is smoothed, the leech makes its body as long and thin as possible. In the space left between the leech's body and the inside wall of the cocoon, it secretes a nutritious fluid similar to egg white. The fertilized eggs are laid in this fluid. The front part of the leech's body then begins to back slowly out of the cocoon, which remains attached to the protected surface of the rock or log. As the leech's head passes through the front and back openings of the cocoon, it seals the openings with **mucus** plugs. These plugs contain the bacteria necessary for the young leeches' digestive processes.

At this stage the cocoon is a soft, clear, colorless bag. The leech presses it flat, and in a few days the cocoon becomes hard and turns dark brown. The parent's job is then finished. When the hatchlings emerge, they are on their own. To stay alive, they must find food and avoid being eaten by larger leeches, insects, birds, or fish.

Leeches that lay their cocoons on land form them in similar ways. They, too, attach their cocoons to rocks, logs, or other protected surfaces. However, a land leech does not press the cocoon flat, but allows it to keep its rounded shape. The land-laid cocoon wall has two layers—a smooth inner layer and a rough, spongy outer layer. Scientists believe that the spongy outer layer, made of a material similar to insect silk, helps protect against water loss.

A few types of leeches produce very thin-walled cocoons and stay near them for many days while the young leeches develop special attachment knobs on the front part of their bodies. Once these knobs are developed, the

The giant Amazon leech with young leeches on its underside. The young will remain attached to the parent until they are ready for their first blood meal.

young break free of the cocoon and attach to the underside of the parent leech. Here they remain to continue their development. Once their suckers are formed, they lose their knob-like organ and hold to the parent by their rear sucker.

Scientists have counted as many as ninety-

34

two young leeches clinging to a parent leech. They remain attached for weeks or months, gaining shelter from the parent and growing stronger. These young leeches do not live if they are taken from the parent too early because they require the gentle, waving movement to provide fresh water and oxygen.

Some leeches, such as the Amazon leech, form cocoons that remain attached to the underside of the parent leech's body. When the hatchlings emerge from the cocoon, they cling to the parent for several weeks until they are ready to eat.

First Blood Meal

The time between hatching and the first blood meal varies widely. Some species can wait four months before they eat, while others can wait only fourteen days, depending on how much yolk they have stored in their body. When the yolk is used up, they must eat.

After the first feeding, the leech usually looks for a dark, quiet place such as the underside of a rock or a log for shelter while it slowly digests its meal.

Each meal enables the leech to grow and mature. As it grows, it can store more blood, so that it can survive longer between each meal. After the second feeding, the male reproductive system matures. Usually by the time a leech has eaten four meals, the female reproductive system has matured also, and the leech reproduces.

Leeches normally live from two to six years and reproduce once or twice a year. In the laboratory, however, exceptions do occur. There leeches are fed on a regular schedule and eat more than they usually would in the wild. One captive Amazon leech laid about two hundred eggs three times a year and produced 750 offspring, all without a mate.

Because there are many different kinds of leeches, these creatures vary greatly in their

36

appearance and behaviors. But no matter what their differences, all leeches draw as little attention to themselves as possible. They feed without being noticed, hide under rocks and logs, and are active mostly at night. In these ways, they have adapted to give themselves the best possible chance for survival. After all, how many people would go where they could see that leeches thrived, or act as a willing host if they knew that a leech had attached to their body?

CHAPTER

~3~

The History of Medicinal Leeching

Two thousand years ago, the practice of medicine was primitive, indeed. People thought evil spirits and poisonous fluids caused disease. Early healers believed the way to rid the body of illness was to bleed the patient.

They used several methods, all of which harmed more than helped. Thorns, animal teeth, pointed sticks and bones, and sharp pieces of flint and shells were used to wound the patient and cause bleeding or bloodletting. Healers in Greece and Malta used tools similar to very small crossbows. Medicine men in areas as far apart as South America and New Guinea used tiny bows and arrows.

Eventually, people learned that these meth-

38

ods caused infection or serious blood loss. They turned to a safer, more precise method of blood-letting—leeching.

The Spread of Leeching

Leeching, the use of these animals to draw blood, was used in the Far East hundreds of years before traders brought knowledge of this method to Europe. Nicander of Colophon, a Greek who lived more than two thousand years ago, is the first known westerner to use them in medical practice.

The belief that evil spirits caused disease faded during the time of Hippocrates (460 to 377 B.C.). It was then that the idea of body humors was developed. People thought the body contained four fluids—blood, phlegm, yellow bile, and black bile—and good health depended on the proper balance of these humors. Leeching continued, but now it was used to balance the humors.

In this illustration from a book published in 1639, a woman applies a leech to her forearm. Other leeches are in a large jar on the table.

For hundreds of years, priests and monks in Europe took care of the sick and injured, often using leeches in their treatments. But in the twelfth century a church council prevented them from continuing the custom of bloodletting. Church leaders felt that some religious houses were taking too much interest in how much

40

they could gain in money or possessions by treating the sick.

Some barbers then began to add these practices to the cutting of hair and trimming of beards. These barbers, called barber-surgeons, not only used leeches, but used other forms of bloodletting as well. Sometimes they would use specially designed tools to cut arteries or veins to drain blood. Unfortunately, these untrained practitioners would sometimes accidentally cut through the vessel or injure a nerve or tendon, causing serious injury. Barber-surgeons also treated wounds and even amputated arms and legs. After bleeding sessions, they would wrap the bloody bandages on a post outside their shops—a sign that showed the barber did surgery as well as haircutting. So was born the barber's symbol, a pole painted with spiraling red and white bands.

For a long time, educated physicians did not perform surgery, leaving that practice to the barber-surgeons. During the sixteenth century, though, a law was passed in England that separated the duties of barbers and surgeons. Barbers stopped practicing surgery and stuck to cutting hair and trimming beards. Physicians took over the treatment of wounds, surgery, and the practice of bloodletting.

During the eighteenth and nineteenth centuries, leeching and other forms of bloodletting were practiced more widely in Europe, especially in France, which was considered the center of medical learning. François Victor Joseph Broussais, an important French physician, thought disease was caused by the inflammation, or swelling, of the digestive tract. Bloodletting, especially leeching, was his main treatment. His standard therapy was to withhold food from the patient and apply at least thirty leeches over the entire body.

Other physicians throughout Europe copied his treatment, causing an enormous demand

During the nineteenth century, pharmacists and physicians kept leeches in decorative jars with tiny air holes in the lids.

for leeches. During this time as many as 100 million leeches per year were used in France. The emperor Napoleon imported nearly 6 million leeches in one year to treat his soldiers. Because of this huge demand, the price for leeches skyrocketed. In France in 1806, one franc bought eighty leeches. Fifteen years later, the same amount of money bought only four of them.

Perhaps because leeches were so expensive, some pharmacies began to rent them. Customers used them to drain stagnating blood from black eyes, to treat **varicose veins**, chest pains, headaches, or because they thought periodic bleeding was healthy.

When the leech had finished its meal, it dropped off. The borrower returned it to the pharmacy, and it was rented by another customer. This practice was seriously flawed because it could spread disease and infection from one customer to another.

Medicinal leeches were in such high demand at this time that they nearly became extinct. England and Russia passed game laws to protect them. Export regulations were established to restrict the size of leeches that could be sold. Resourceful people began raising leeches, and leech farms flourished.

In the past, women and children were paid to wade bare-legged into ponds to gather leeches for use in medicine.

Leeching in America

In America, the craze for bloodletting, particularly leeching, never reached the fever pitch that it did in Europe. Still, 1.5 million leeches were used each year in the United States during the fifty years following the Civil War. In America, the New World species of medicinal leech,

Macrobdella decora, was often used for leech-ing. But the traditional European variety, *Hirudo medicinalis*, considered the best leech availa-ble, was imported in large quantities.

There were many ways to apply leeches. One was to remove the leech from the water and dry it in a soft towel for thirty minutes before use. The skin where the leech was to be applied was sponged with warm water to increase blood flow to the area. The leech then was placed on the skin. If it did not take hold immediately, milk or milk with sugar was ap-plied to the skin to encourage the leech to bite. If several leeches were needed, they were placed in a teacup or a wine glass, which was then turned upside down over the area. This method kept them in the right place until they attached.

Usually, leeches simply dropped off when they were full. If not, salt, snuff, or vinegar was applied. Bleeding, which would have continued for several hours, was halted by applying pressure, or a drop of perchloride of iron, or by **suturing** (stitching), or **cauterizing** (burning) with a red-hot needle.

Leeching of Famous Figures

Bloodletting was a practice accepted by ordinary citizens and famous people alike. Julius Caesar, the Duke of Wellington, Leo Tolstoy, and Napoleon were leeched. George Washington was bled heavily by Dr. Benjamin Rush in 1799 while being treated for laryngitis. Washington died about twenty-four hours after that "treatment."

By 1885, microorganisms became known as the cause of disease, and different medical treatments were invented. Bloodletting, a "healing" procedure several thousand years old, fell from favor.

Still, tradition dies hard. In the 1920s, American physicians still used bloodletting to treat pneumonia. In 1941, when Adolf Hitler complained of ringing ears, his doctor applied leeches. They were placed behind his ears, and it is said that the ringing stopped. In 1953, Soviet doctors leeched Joseph Stalin shortly before he died. Even now, a few pharmacies in the United States and abroad sell leeches over-the-counter to customers who use them to relieve the pressure of black eyes, varicose veins, and bruises.

The use of leeches to draw blood, like many other folk remedies from earlier times, does have a proper place in modern-day medicine. Although overused as the ancient remedy for whatever ailed people, physicians now have rediscovered leeching as a highly effective modern treatment for special medical problems.

~4~

Science Latches onto Leeches

Today the leech is acquiring a new role as an important tool in medicine and scientific research. Doctors at some hospitals in the United States and in Europe use the medicinal leech to help them reattach fingers, ears, or other body parts that have been severed or badly damaged in accidents.

Surgeons using microscopes can restore circulation to reattached parts by sewing together tiny nerves and arteries. They use sutures so fine that they are barely visible to the naked eye. Often, though, veins are too delicate to reconnect. As a result, fresh blood flows into the finger or other body part through the repaired arteries, but there is no route for the return.

Blood pools in the reattached part, which causes swelling and pain. The pressure can cause clots to form in the arteries, stopping the flow of blood and preventing cells in the re-attached part from getting oxygen and nour-ishment. If this condition continues, the re-attached body part will be lost. To save it until the damaged blood vessels can grow and re-store proper circulation, doctors usually make a small cut so it bleeds and relieves the pressure buildup. Yet repeated pricks can damage the tissue, and they do not always work. Sometimes blood transfusions have to be given to make up for the large blood loss which can result from pricking.

Leeches in Modern Medicine

Some doctors instead have begun to use leeches to drain off the unwanted blood. Not only is the procedure nearly pain-free because of the anesthetic in the leech saliva, but it is

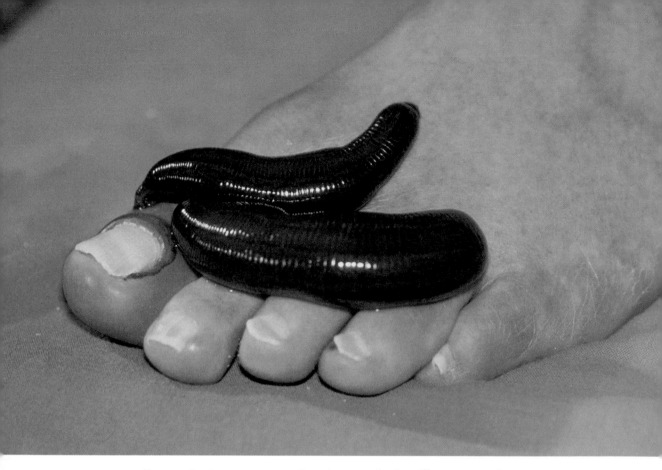

Some doctors now use leeches to drain off unwanted blood in an injured or reattached body part, such as this bruised and swollen big toe.

also effective. Because of the spreading factor and the anticoagulant, **hirudin**, also in leech saliva, the wound continues to bleed for three to five hours after the leech drops off. This keeps the swelling down for an extended time. Between two and ten leeches are applied throughout the day. This treatment may have to continue for up to nine days until new veins can grow.

Once a leech has been used on a patient, it is not used again to prevent any chance of spreading infection from one patient to another. Hospitals sometimes return used leeches to the leech supplier, where they can be used to breed more leeches. Some hospitals choose instead to burn used leeches after first deadening the leeches' senses. They do this by placing the leeches in a weak alcohol solution.

One boy who has been helped by leeches is Guy Condelli. In 1985, at the age of five, he was attacked by a dog which bit off his ear. Guy was rushed to Boston Children's Hospital. There, Dr. Joseph Upton reattached his ear in a delicate operation. Afterward, blood was flowing to his ear, but clots were forming in the veins. The blood could not return from the ear to Guy's heart. Anticoagulants injected into his circulatory system failed to solve the problem.

52

Dr. Upton remembered using leeches successfully when he was a surgeon in Vietnam. He tried them again on Guy's ear. Two 1.5-inch (3.8-centimeter) leeches were applied. At 8 inches (20.3 centimeters) long, the leeches looked like blood-filled cigars when they dropped off. During the following week, two dozen more leeches were applied before Guy went home with his ear safely reattached.

People who are concerned about the medical use of leeches are pleased when they see that they work, and work painlessly. The mother of another five-year-old boy said she got chills the first few times leeches were placed on her son's repaired hand. Her son had lost two fingers in an accident that happened while he was playing in the family's garage. The boy, though, didn't mind the leeches at all. In fact, he named one "Larry the Leech."

Another case involved a cabinetmaker who caught her fingers in a power saw. Her index

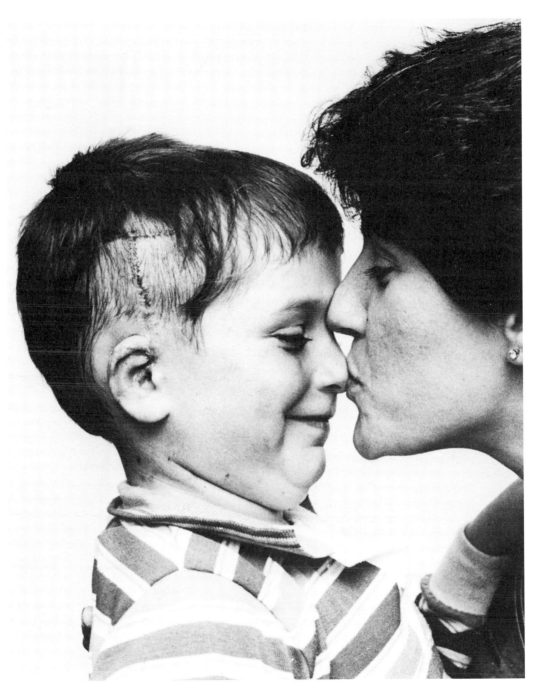

Guy Condelli and his mother are pleased that surgeons,
aided by leeches, were able to save his reattached ear.

finger was severed and three others badly damaged. Surgeons reattached the finger and repaired the others. But the reattached finger began to turn black as clots formed in the veins. Doctors feared they would have to remove the reattached finger.

Instead, they tried leeches. The first leech was not hungry and could not be coaxed to feed. The second, however, began to feed immediately. Within ten minutes the cabinetmaker's finger was a healthy pink once again. The leech stayed on for two hours and then dropped off and went to sleep. Doctors applied one other leech, which stayed on for twenty-four hours. At first the cabinetmaker was worried that the leeches would hurt or sting, but the pain in her swollen hand actually decreased when the leeches fed. Today, she still has all her fingers.

At present, European doctors use leeches in more medical procedures than American physicians. In England, for example, leeches are used to decrease the swelling of black eyes, allowing them to be opened for observation. In one case, leeches were applied after microsurgeons re-attached a British steelworker's scalp. It was torn off when his shoulder-length hair was caught in machinery in a factory accident. In France, leeches are used not only to reattach fingers, but also in skin grafts—operations in which large sections of skin, nerves, and blood vessels are moved from one part of the body to another.

Leeches in Scientific Research

In the United States, though, some research-ers have made leeches a focal part of their work. At the Temple University School of Medi-cine in Philadelphia, scientists have isolated a protein from the **salivary glands** of the giant Amazon leech, *Haementeria ghilianii*. They have found that the saliva contains an exciting

blend of anesthetics, anticoagulants, and antibi-
otics, including a previously unknown protein
called **hementin**. The researchers believe that
hementin will be important in treating diseases
that cause heart attacks and strokes, and in dis-
solving blood clots in major blood vessels.
Hementin could be injected into or near the
clot, concentrating the treatment where it is
needed.

Other researchers at Pennsylvania Hospital's
Laboratory of Experimental Oncology in Philadelphia are working with another, and as yet
unnamed, extract from the leech salivary gland.
Their studies of lung cancer in mice show that
the extract is able to inhibit, or slow, **metastases**—the spread of cancer from one part of
the body to another. The extract also reduces
the undesirable side effects of some cancer
drugs and strengthens the natural immune system of the body—the system that acts to protect people from disease.

A researcher at the University of California at Berkeley ▶
uses a powerful microscope to study the nervous system
of the leech.

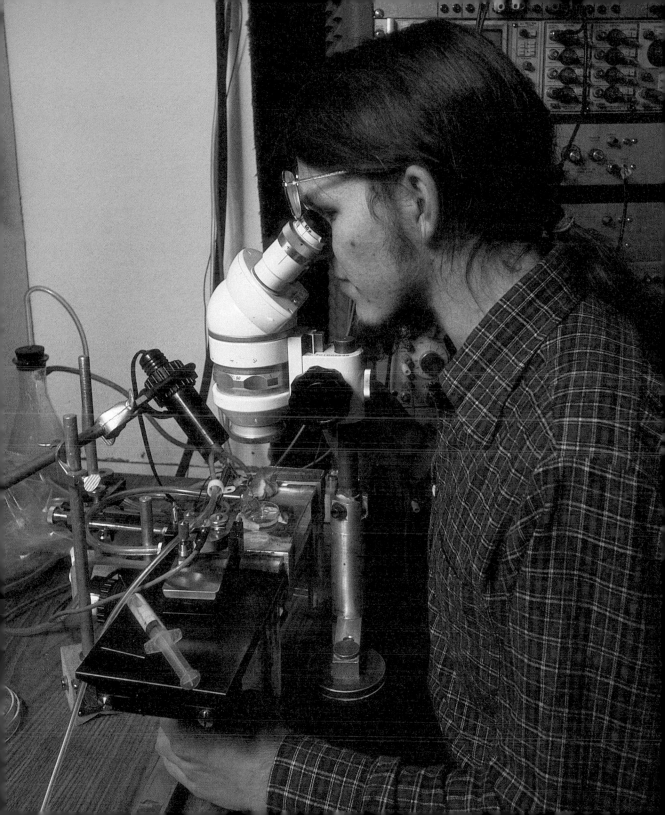

58

The discovery of these extracts opens up an exciting field of research that may produce a number of medically useful substances.

Leeches have also become important in other areas of scientific research. Because of the very large and simple nervous system of the leech, it is an ideal animal to use in laboratory studies of nerve cells. Higher life forms, including humans, have millions of **neurons**, or nerve cells, in their brains. Leeches have only ten thousand in their entire nervous system.

The leech's nervous system is made up of a central nerve cord with a group of nerve cells called a **ganglion** at the head and tail. These function as the leech's "brains." In addition, each segment has a smaller ganglion that contains about two hundred nerve cells. Not only are the nerve cells large, but they are spread out in a single layer. That makes it possible to observe each cell individually under the microscope.

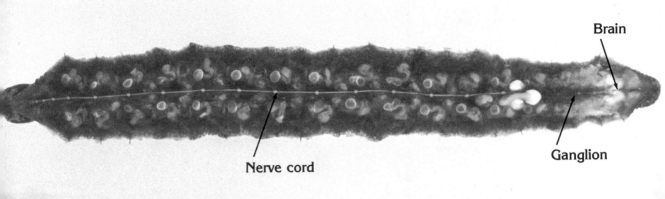

Brain

Ganglion

Nerve cord

The main parts of the leech's nervous system can be seen in this photograph.

Scientists have learned how to insert a small tool called a **microelectrode** into single neurons to stimulate them and measure individual responses from these neurons. In many cases it is even possible to identify the function that each nerve cell has for the animal. For example, some of the cells are motor neurons that cause specific muscles in the leech's body to move. Other neurons are sensory neurons that react to light, heat, and vibrations.

Using the microelectrodes, scientists have charted connections between nerve cells. They have discovered the exact sequence, or order, in which individual nerve cells fire to produce swimming movements in leeches. So far researchers have not been able to do this with any other animal. Yet in nature, the way in which one species works is often repeated in other species. Learning the workings of a simple animal such as the leech may provide clues to the workings of more complex species, including humans.

Scientists believe the discoveries will help explain how brain cells respond to drugs, and how learning and memory work. They also hope the leech will provide information about human learning problems and birth defects so that one day we can prevent them.

Recent discoveries that have been made using leeches have led scientists to predict that these creatures soon will be used routinely in behavioral and medical research. One scientist in particular has played a major role in providing leeches for modern research. In fact, some kinds of leeches would not be available at all without the adventures and efforts of Dr. Roy Sawyer.

CHAPTER

~5~

Braving the Wilds for Science

Roy Sawyer, a young American scientist, was knee-deep in a humid marsh in French Guiana, a country in northern South America. Suddenly, in front of him, a large snakelike animal slithered through the water. Observing it closely to make certain it was not a snake, Sawyer quickly grabbed it with his hand. He pulled a huge, greenish-brown, bloodsucking worm out of the water, glanced at it quickly, and then flung it to the grass at the edge of the marsh.

Scrambling to the bank, Sawyer picked up the leech again, and suddenly noticed small leech hatchlings attaching themselves to his hands.

"Help me get them off before they begin to

feed," he said with a note of urgency. Other members of the expedition from the University of California at Berkeley quickly lifted off the young leeches and dropped them into plastic collecting bags. More than 180 of the hatchlings, looking for their first blood meal, were captured.

Dr. Sawyer had just found an Amazon leech, *Haementeria ghilianii*—the world's largest leech—which can grow to 18 inches (45.7 centimeters) and, some say, can live as long as twenty years.

"Grandma Moses"

Haementeria was first discovered in 1849 by a naturalist working near the flood plains of the Amazon River. The species was spotted again in 1893 in French Guiana, duly recorded in scientific writings, and then forgotten.

But Sawyer, one of the world's few leech experts, discovered the little-known information and obtained two live specimens from a scientist working in South America. One died almost immediately; the other, nicknamed "Grandma Moses," produced more than 750 offspring in just three years. Now it is preserved as part of the permanent collection at the Smithsonian Institution in Washington, D.C.

Sawyer, realizing *Haementeria*'s importance to medical and biological science, wanted to start a breeding colony at the University of California at Berkeley. He and his staff had managed to keep Grandma Moses alive only through a process of trial and error. They knew that more precise information about feeding habits, water temperature, and other data had to be obtained if they were to grow leeches over a long period of time. To gather this data, Sawyer organized the expedition to South America. He wanted not only to collect more leeches, but to learn as much as possible about *Haementeria*'s natural habitat.

To learn about leeches in their natural habitat, expedition ▶ members string a fifty-meter measuring rope tied in knots every meter. They chart the type and amount of vegetation, water temperature, and depth of water.

66

In Search of the World's Largest Leech

The six-person expedition team began its search for the giant leech in Surinam, a country bordering French Guiana. The area is home to flesh-eating piranhas, vampire bats, some of the world's largest scorpions and tarantulas, and electric eels with a charge powerful enough to topple a horse. Even more dangerous were snakes, the fer-de-lance and the bushmaster, which are the largest poisonous snakes in the Americas. Despite the animal hazards, according to Sawyer, the greatest danger was getting lost in this nearly uninhabited corner of the continent.

When the team failed to find *Haementeria* in the swamps of Surinam, they headed for French Guiana. They stayed at the village of Trou Poisson, in a former schoolhouse now used as a hotel for French scientists studying malaria. In Guiana's primitive back country, the schoolhouse's main advantages were the bath-

*Leech expedition members from the University of Cali-
fornia at Berkeley unload equipment from their Land
Rover in French Guiana.*

room and the screens on the windows. The
screens kept out vampire bats and the hordes
of mosquitoes which rose from the stagnant
marshes at dusk.

The day after their arrival, the expedition
found what it sought—leeches. The team went
to the same stagnant pond where Grandma

68

Moses had been discovered two years earlier. Expedition members, wearing rubber waders, slogged into the murky water. Roy Sawyer waded out from the bank, raised one boot out of the water, and spied six greenish-brown hatchlings wriggling across his boot.

The next day they returned to the same spot with a young Haitian cattle-tender named Jean-Claude Louis. He often went into the swamps without boots. Expedition members thought that leeches would be more attracted to Jean-Claude's skin than to their rubber waders.

They were right. Jean-Claude knew that leeches attach themselves rather quickly but do not begin to feed until several minutes later. Wearing only swimming trunks, he waded out to waist-deep water. He then returned with several dozen leeches clinging to his legs and feet. He quickly picked them off, but he kept a lime which, when sliced and rubbed on a leech, causes it to detach instantly. The lime would be

An expedition member observes leech hatchlings crawling across his rubber waders.

used only in an emergency since lime juice also kills leeches.

By midday 100 leeches were collected, and expedition members had gathered much important data. They discovered that leeches were most likely to gather in shaded, weedy areas of the marsh. They also learned that they had unfortunately arrived during breeding season. During that time adult leeches lose their appetite for blood and burrow into the mud. As a result, only one very large leech was found. The scientists named the 13-inch (33-centimeter) *Haementeria* "Christine."

As their leech collection techniques improved, the expedition spent a few more days in Guiana, and then packed for home. Sawyer and his team placed Christine, the young leeches, and the hatchlings in plastic bags filled with moist leaves. They placed these in airtight cracker tins which were shipped back to the United States by air freight. Although they were

At the Berkeley laboratory, researchers created an artificial ▶ *pond in which the conditions were as close as possible to the natural environment of the Amazon leech.*

lost in transit for more than a week, the leeches arrived in excellent condition.

   With those leeches, Dr. Sawyer produced a colony of *Haementeria* at the University of California at Berkeley. Once his research at Berkeley was completed, he returned to the University College of Swansea in Swansea, Wales, with offspring from the Berkeley colony. Swansea was where his wife's family lived, and it was also where he had earned his doctoral degree in zoology. Now he operates a leech farm in Swansea and also serves on the faculty of the university. The farm supplies leeches and extracts from leech saliva to medical and scientific communities around the world.

CHAPTER

~6~

A Visit to a Leech Farm

Imagine what a leech farm looks like. Do you think of big swamps in the open? Or do you see a wet jungle area enclosed by fences? Maybe you even imagine a big red barn full of leeches instead of cows, horses, or pigs.

The only self-sufficient leech farm in the world—Biopharm—doesn't look like any of those scenes. Leeches are raised in a former steel mill in the industrial district of Swansea, Wales. In the parking lot next to the brownish-grey building, lengths of iron pipe and rusting metal lie in scattered heaps. Railroad tracks border the entrance, and the River Tawe flows at the edge of the parking lot.

The leech farm is hard to find. No signs

point the way, and there is no name on the entry door. The door is at the top of a flight of metal stairs at the back of a commercial park where other businesses are located. Workers in the other businesses would probably be surprised to know they were working next to one of the largest concentrations of leeches anywhere. It is also the source of most leeches used in medicines and research worldwide.

Life on the Farm

Biopharm was started in 1984 by Dr. Roy Sawyer. He and about ten employees are breeding not only giant Amazon leeches like "Grandma Moses," but the medicinal leech and five other species as well. In a series of windowless offices, laboratories, storage areas, and warm and cold rooms, leeches are raised, maintained, and packed for shipment to nearly all parts of the world.

Perhaps as many as two-thirds of the hospitals in Great Britain and Europe use leeches. Still, the hospitals in the United States that use leeches order the largest quantities—from 500 to 1,200 leeches per shipment—and their use is growing.

When leeches are needed, Biopharm moves fast. Leeches can arrive ready for action at any hospital in Great Britain within five to six hours. They can arrive within twenty-four hours at hospitals in the United States.

Normally, shipments of leeches are sent by train from Swansea to London and then routed by airplane or another train to their final destination. But in cases of great urgency, Lorna Sawyer, Dr. Sawyer's wife, drives the shipment to London's Heathrow Airport, three hours from Swansea, and loads the leeches on the next flight out.

Currently, Biopharm has more than 100,000 leeches in what Sawyer and his helpers call a "cold room." The leeches thrive in large vats,

A researcher at Biopharm holds medicinal leeches that are kept in tanks in a cold room. Notice the velcro strips along

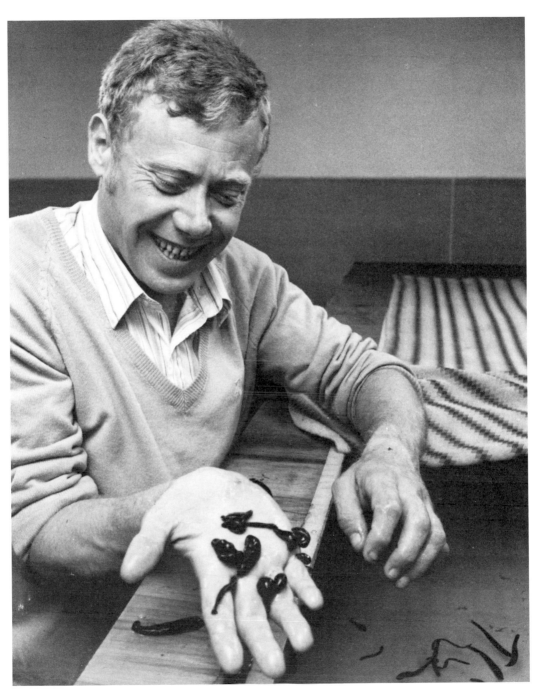

*the edges of the tank and the cloth covers, which are
pulled back to show the leeches.*

tubs, fish tanks, and even a fiberglass dinghy, all connected by tubes and pipes. Each container is covered securely with cloth, which is held tightly around the vats with velcro strips. This keeps the leeches from escaping, which they will do given the slightest chance. The floors at Biopharm are bare concrete and vinyl tile. If leeches escape, bare floors give them no place to hide. Dr. Sawyer and his wife once had to pull up the wall-to-wall carpeting in their living room to locate a pair of escaped Chinese leeches he had brought home.

Most of the leeches live in the cold room because they are **dormant** in cool temperatures and easier to handle and maintain. Cold slows their bodily activities and reduces their food intake. In this condition they are slow to attach. Cold-room leeches are kept hungry so that when they arrive at hospitals, they will attach quickly to a patient and feed without delay.

"Hungry" leeches are divided into "teams."

Each team is fed about once every twelve months on a rotating basis. Once fed, they won't eat again for about a year. Leeches at Biopharm feed on a special ox-blood sausage prepared in the laboratory. Hundreds are fed at one time. Leeches recently nourished are separated from hungry leeches because the hungry ones will feed on those that are well fed.

In addition to the cold room, Biopharm has two warm rooms containing about ten thousand leeches. Here leeches are kept in different stages of growth. Some are breeding, others are being fattened for breeding, and the rest are hatchlings.

Breeding Leeches

Breeding chambers in the warm room hold from fifteen to thirty medicinal leeches per container, and these leeches produce fifty to eighty new cocoons each week. Dr. Sawyer and his associates are trying to increase that number because of the heightened demand for leeches. Each cocoon produces fifteen to twenty hatchlings for a total production of about ten to fifteen hundred hatchlings weekly.

During breeding, leeches are not handled or disturbed in any way, nor are cocoons moved or touched. The slightest disturbance, such as stagnant water or temperature changes, will cause breeding leeches to fail to produce hatchlings.

Researchers at Biopharm spend about two months fattening up teams of leeches for breeding. Only healthy, well-fed leeches go into the breeding chamber. Those that are undersized are reserved for research purposes.

One team of medicinal leeches became so large that workers nicknamed it the "A Team" after the popular television show. Most were 8 to 9 inches (20.5 to 23 centimeters) long. Researchers believe it is possible that medicinal leeches could reach the size of giant Amazon leeches, given enough food and time.

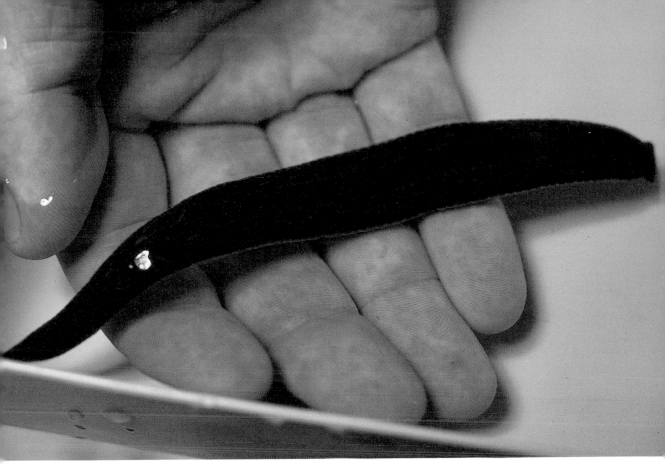

Hirudo medicinalis is the most commonly used leech in laboratories.

Amazon leeches are more temperamental than medicinal leeches. If they are not always kept in one of the warm rooms, they will not breed. During breeding, Amazon leeches produce from 200 to 300 young, each valued at about $150, or nearly $40,000 for the offspring of just one giant leech.

82

In 1979, Dr. Sawyer had about ten thousand Amazon leeches, which he had raised from the thirty-five he captured in South America. Disaster struck one weekend when the heating system in the laboratory broke. All but a handful of the leeches died.

Each of the Amazon leeches now at Biopharm and those in research centers around the world are the offspring of the few leeches Dr. Sawyer was able to salvage. In order to prevent a similar situation from occurring, one-third of the Amazon leeches are kept in the breeding room, and the others are kept in a warm room. The breeding room has been specially insulated so that even without power it will maintain warmth for twenty-four to thirty-six hours.

"Milking" Leeches for Science

Along with the warm rooms and the cold room, Biopharm has a biochemistry room where the chief biochemist "milks" the leeches to get their salivary extracts and other chemicals used in research. The process he uses is known only to him.

The biochemist makes a gel to which he adds the salivary extracts. He places the mixture in a fraction collector, a special machine that separates the anticoagulants, enzymes, and other chemical components. Once they are separated, he can freeze-dry them, bottle and store them in a refrigerator, or combine them in any number of ways.

What he collects is used in Biopharm's research or is sold to other researchers around the world. Material is shipped in small glass bottles similar to the injection vials physicians use. Leech extracts are in short supply because thousands of leeches must be sacrificed to get even small quantities of the salivary extracts in pure form. That is why Biopharm raises such large numbers of leeches.

Researchers at Biopharm are studying ways

to produce these chemicals in unlimited quantities through genetic engineering. They are
working to isolate specific genes—tiny elements
that transmit particular traits—such as the
hementin gene, from the leech salivary extracts.
They plan to inject these genes into yeast,
single-celled organisms that can be grown
quickly in a laboratory. When injected with a
foreign gene such as hementin, the yeast will
produce that gene as it multiplies. This process
would decrease the need for so many leeches
and make more research possible with an
expanded supply of leech extracts.

Perhaps in the future scientists and doctors
will not use leeches, but only the extracts from
their salivary glands. Still, whether the whole
leech or just a part of it is used, the leech's
place in science and medicine has been well
established.

CHAPTER

~7~

Collecting and Keeping Leeches

Finding leeches is not high on the list of favorite activities for most people. In fact, more often than not, leeches find *them*. Even so, collecting and studying leeches as a science project can be a valuable learning experience for young people when they work with an adult. Keeping leeches at home as pets, though, is not recommended.

Collecting Leeches

Leeches can be found in nearly every place where water is available. The most likely places are the shallows of ponds, lakes, or marshes that are rich in plant life and snails. The best time of year to find leeches is spring or summer.

86

Most leeches are **nocturnal**, which means they are active at night and thus avoid light. You can find them in shady areas of ponds or in dark places under rocks, logs, plants, and debris on lake bottoms. Also, they attach easily to smooth surfaces of litter such as glass bottles or jars, metal cans, or plastic objects. Some species of leeches attach to fish, frogs, turtles, or other small hosts, though most drop off after they have fed.

Once found, they can be collected very gently with **forceps**, large, tweezerlike instruments. Some of the active swimming species can be collected at night with a small dip net.

The majority of leeches you find probably will not be bloodsucking varieties. Instead, they are predators that eat small **invertebrates** such as snails, worms, and insect larvae. Other varieties eat fish eggs and amphibian eggs, or feast on decaying matter at the bottom of lakes or ponds. Many of the bloodsucking leeches attach

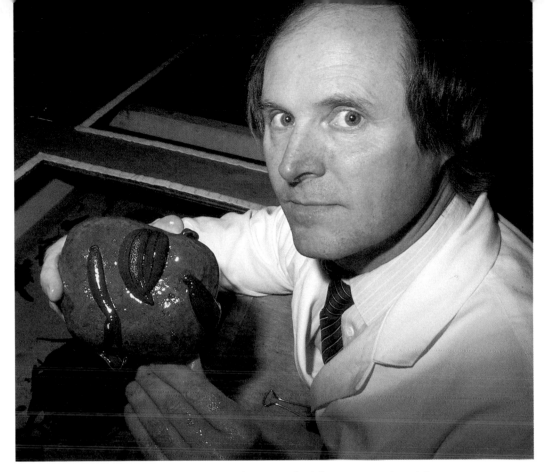

At Biopharm, Dr. Roy Sawyer holds up a rock with leeches attached to it.

only to specific animals such as turtles, fish, frogs, or birds. In favorable environments, as many as seven hundred leeches per square meter have been found.

If you don't find leeches on your first try, keep looking because there are plenty of leeches around to capture. And remember, always be alert when looking for leeches. When their usual food source is not available, some species will feed on whatever is available, even people.

Over the years, different ways to discourage leeches have been developed. Common anti-leech remedies are insect repellent or lemon juice squirted over feet and ankles to discourage leeches from attaching. Credit cards work well as tools for flicking the pests off before they can attach. However, you must always watch for the arrival of each new leech.

If these measures fail and a leech does attach, it is not a good idea to try to pull it off with force. Because of the leech's strong sucking power, a part of the leech could be left behind and later cause infection.

Gently squeezing the leech will sometimes cause it to let go. Leech scientists in the field often carry lemons or limes to slice and squeeze over attached leeches. The juice causes leeches to detach almost immediately, but usually kills them as well.

Maintaining Leeches

Keeping leeches in a laboratory or classroom is not difficult. The main requirement is water that does not contain harmful substances such as chlorine, copper, or other chemicals. Water from the place where the leeches were collected, rainwater, or bottled spring water works well. Tap water is usually suitable if allowed to stand in an open container until all of the chlorine escapes—for at least two full days. Distilled water should not be used because it will harm the animals' mineral balance.

The water must be clean and cool. As soon as it shows signs of becoming dirty, it should be changed. Care must be taken, however, to avoid sudden temperature changes, and to keep leeches out of direct sunlight or artificial light sources.

Most any type of glass or plastic container such as an aquarium can be used to hold leeches, but it must be securely covered. Some species of leeches are amphibious and like to crawl and explore. If you use a perforated lid, the holes must be very tiny. Leeches can lengthen and thin their bodies to fit through remarkably small openings. Cloth covers, tightly secured with string, rubber bands, tape, or velcro, are probably safer. Cloth allows air to circulate into the container but prevents leeches from escaping.

Leeches do not like to be crowded. Depending on the size of leech, not more than fifteen to twenty-five leeches should be placed in a 5-gallon (19-liter) container. Since some species like to attach to the sides above the water line, leave from 5 to 6 inches (12 to 15 centimeters) of air space between the water and the top of the container. A container with a sloping end can also allow leeches to crawl out of the water.

Leeches may be kept in a securely covered glass container with air space between the water and the top of the container.

You can make one by placing several inches of water in an aquarium and adding a sand bank that rises well above the water. Shells, pebbles, small sticks, and aquatic plants can be placed in the container. These provide places for the leeches to hide, although they are not necessary.

Feeding Leeches

Leeches do not eat often. If you keep them only a few weeks, you probably will not need to feed them. Leeches kept for longer periods, though, will need to eat. Because food requirements are different for each type of leech, knowing what to provide for the leeches you capture can be difficult. In general, small predator leeches will feed on live snails and slugs. Larger, bloodsucking leeches will require a meal from an animal such as a turtle or frog every month or two. Earthworms, frog eggs, the **larvae**, or young of insects, and even raw ground meat can be given to other bloodsucking species at six-month intervals.

Experiments with Leeches*

Having captured some leeches, you may want to perform a few simple experiments with them. The following suggestions for experiments demonstrate leeches' abilities to find

*Adapted from "Leeches: New Roles for an Old Medicine," by Dr. Roy T. Sawyer, from *Bulletin*, Spring 1986. Reprinted with permission, Ward's Natural Science Establishment, Inc.

food. Maybe you can think of other experiments as well.

Leeches and Vibrations

Place a live leech in a 16-ounce (480-milliliter) container filled with pond water. Allow the leech to remain undisturbed for five minutes. Then use a plastic ruler, or similar flat object, to stir up the water surface.

Leeches are extremely sensitive to vibrations in the water and attempt to discover their source. The leech's segmental receptors—the small colored dots, spots, or patches on its body segments—are what enable it to detect the water movement. These receptors, though, are not easily seen.

Leeches and Light

Leeches are very sensitive to light or shadows passing over them. Place a leech in a 16-ounce clear container. Allow it to remain undis-

turbed for a few minutes in a well-lighted room. Then pass your hand over the container so that a shadow falls on the leech. It should immediately become excited and start swinging its head.

The leech's two types of light receptors, eyespots (ocelli) and eyes (groups of ocelli in a pigment cup located in the head region) react to changing light conditions. You will not be able to see eyespots since they are single cells. Eyes, though, can be seen. Their location and number are among the features scientists use to identify different types of leeches.

Leeches and Odor Detection

Take a clean, empty 16-ounce jar and make a fingerprint on the *inside* of the container. Be sure no other fingerprints are visible. Mark the fingerprint on the *outside* of the container by circling it with a grease pencil. Add pond water and a leech to the container. Observe the animal

◀ *Stirring the water with a ruler will cause leeches to swim around looking for the source of the disturbance.*

closely. The leech should pass over the area of the fingerprint and act as if it has detected its "odor."

What the leech is detecting are the skin oils you left in the fingerprint. Leeches are keenly receptive to human **sebum** (skin oils), as well as many other substances. If even a small amount of a substance such as blood, liver, or a chemical is placed in the water with the leech, it will detect the substance and react to it.

Try adding a few drops of dilute (watered-down) beef broth to a 16-ounce container of pond water containing a leech. Keep track of the number of drops necessary to excite the leech. You can also try very dilute solutions of acetic acid, vinegar, or other irritants.

Disposal and Preservation of Leeches

Once your study of leeches is complete, you may return the leeches to the place where they were found.

If you want to preserve a leech, it can be dropped into carbonated water, or weak solutions of alcohol or chloroform. The concentration should be gradually increased over a thirty- to sixty-minute period so that the leech will not contract strongly, making its features hard to recognize.

Once a leech no longer responds when touched with forceps, it may be washed and straightened by gently passing it between your fingers. Then keep it flat while dipping it in a fixative such as 2 percent formaldehyde. Once the solution has soaked into the leech, it will remain stiff and can be washed and placed in a jar or vial with an 80 percent alcohol solution for permanent preservation.

A Last Word

As you have read and learned about leeches, some of the mystery and fear surrounding these remarkable creatures may have changed into an

understanding of their place in the natural world. Leeches can be beneficial to people, now and in the future. Perhaps Dr. Roy Sawyer expresses it best when he says, "I think we have proven [through research] that a seemingly insignificant animal, one that is generally regarded as a repulsive pest, has real value. I can't think of a better argument for the preservation of wildlife and all our natural resources."

◀ *Dr. Roy Sawyer shows a young visitor an Amazon leech at Biopharm.*

APPENDIX A

Scientific Classification of Leeches

Just as there are many varieties of dogs, cats, and birds, there are different kinds of leeches. In fact, scientists have identified 720 species of leeches. Leeches are divided into two classes, four major orders or groups, and five families. The four orders are described below:

Acanthobdellida (uh-CAN-thahb-DEHL-uh-dah)
The Acanthobdellida are unusual leeches that live on fish—mostly salmon—in the Soviet Union. They have only one sucker, on the back part of their body. Small, stiff bristles on the sides of their body resemble those on earthworms. Some people call this order the connecting link between leeches and the earthworms.

Rhynchobdellida (RIHN-kahb-DEHL-uh-dah)
The Rhynchobdellida are bloodsucking leeches that use a sharp, needlelike organ called a proboscis, which they insert into their host to draw out blood. *Haementeria ghilianii*, the giant Amazon leech, belongs to this order.
The Rhynchobdellida are divided into two major families. One is found in the ocean, from the poles to the tropics, and feeds on fish and even young penguins. The other lives only in freshwater, feeding mostly on the

blood of fish, frogs, snakes, and larger animals that come into the water to drink or swim. A few live on the body fluids of snails and insect larvae. These leeches are considered predators rather than parasites because they usually kill their host.

Pharyngobdellida (FAIR-an-gahb-DEHL-uh-dah)
The Pharyngobdellida are the nonbloodsucking order of leeches and are sometimes called worm-leeches. They swallow their prey whole and are considered predators. Most often they feed on worms, insect larvae, and small snails.

Some types spend their entire lives in water, while others spend much of their lives out of water. In Europe one type is often dug up in gardens, where they prey on earthworms.

Gnathobdellida (NATH-ahb-DEHL-uh-dah)
Gnathobdellida have jaws and teeth with which they bite their host. There are two major families of Gnathobdellida: one lives in water, the other on land. The famous medicinal leech, *Hirudo medicinalis*, is an aquatic leech found in Europe, south and east Asia, and in North America, where it has been introduced. In Great Britain, it is nearly extinct.

The land leeches in this order are bloodsucking types that attack both humans and animals. They are found on damp vegetation in tropical, rainy areas such as India, Sri Lanka, Vietnam, and other parts of southeast Asia.

Classification of the Hirudinea

Phylum Annelida
    Superclass Clitellata
        Class Oligochaeta (Earthworms)
        Class Hirudinea (Leeches)
            Order Acanthobdellida
            Order Rhynchobdellida
                Family Piscicolidae
                Family Glossiphoniidae
            Order Pharyngobdellida
                Family Erpobdellidae
            Order Gnathobdellida
                Family Hirudinidae
                Family Haemadipsidae

APPENDIX B

Leech Suppliers and Sources for Laboratory Aids

People not inclined to collect leeches in the wild may order them from biological supply houses. Most suppliers will ship live leeches only to hospitals, research centers, or to classroom science teachers rather than to individual students.

Information on current prices, shipping arrangements, and other details may be obtained by contacting the following companies:

Biopharm U.S.A., Ltd.
701 East Bay Street
Box 1212
Charleston, South Carolina 29403
(803) 742-3537

Carolina Biological Supply Company
P.O. Box 1059
2700 York Road
Burlington, North Carolina 27215
Toll Free 1-800-334-5551
North Carolina customers call 1-800-632-1231

Carolina Biological Supply Company
Powell Laboratories Division
P.O. Box 187
Gladstone, Oregon 97027
Toll Free 1-800-547-1733
(503) 656-1641

Connecticut Valley Biological Supply Company
82 Valley Road
Southampton, Massachusetts 01073
Toll Free 1-800-628-7748
(413) 527-4030

Leeches (USA) Limited
300 Shames Avenue
Westbury, New York 11590
(516) 333-2570

Ward's Natural Science Establishment, Inc.
5100 West Henrietta Road
P.O. Box 92912
Rochester, New York 14692-9012
1-800-962-2660

GLOSSARY

amphibious (am-FIHB-ee-uhs)—able to live both on land and in water

anesthetic (an-ehs-THEHT-ihk)—a substance that numbs or causes a loss of feeling or pain in the body; used in surgery to cause loss of consciousness

annelids (AN-uh-lihds)—worms with bodies made of joined segments or rings; earthworms and leeches are both annelids

annuli (AN-yoo-lee)—structures or markings resembling rings

cauterizing (KAWT-uh-rize-ing)—burning with a hot iron or needle to close a wound; similar to branding

clitellum (kly-TEHL-uhm)—a band around the leech from which the cocoon is secreted

coagulating (koh-AG-yoo-layt-ing)—forming into a thick mass; clotting

cocoon (kuh-KOON)—the protective covering produced by leeches and some insect larvae for protection during an inactive stage of growth

dormant (DOHR-mant)—being in an inactive state, not moving or growing for a period of time; some animals and plants are dormant during the winter

embryo (EHM-bree-o)—the earliest stage of growth of an animal before it is born or hatched

enzymes (EHN-zymz)—substances produced by living cells that cause or speed up reactions in the body without causing change to themselves

extinct (ek-STINKT)—no longer living anywhere on earth

forceps (FOHR-sehps)—small tools such as tongs or pincers for gripping and holding things

ganglion (GAN-glee-uhn)—a mass of nerve cells usually outside the brain or spinal cord; it serves as a center from which nerve impulses are sent

hementin (hee-MEHN-tihn)—a protein secreted by the salivary glands of *Haementeria ghilianii*; it is able to dissolve blood clots after they form

hermaphrodites (hur-MAF-roh-dyts)—animals having both male and female reproductive organs

hirudin (HIH-roo-dihn or **hih-ROO-dihn)**—the most powerful, natural anticoagulant known; it is found in the salivary gland of the medicinal leech, *Hirudo medicinalis*

invertebrate (ihn-VUR-tuh-breht)—an animal that does not have a backbone, including sponges, worms, leeches, lobsters, and insects

larvae (LAHR-vee)—the early form of an animal or insect that changes when it becomes an adult; this is often a wormlike stage

metastases (muh-TAYS-tuh-seez)—the spread of disease from one part of the body to another; cancer cells often spread to other parts of the body by way of the bloodstream

microelectrode (MY-kroh-eh-LEHK-trohd)—a very tiny tool used to track the flow of electrical current through a cell

mucus (MYOO-kuhs)—a slimy, slippery material produced by leeches and other animals to moisten and protect areas of the body exposed to the air

neurons (NU-rahnz)—nerve cells, which are the basic working unit of the nervous system

nocturnal (nahk·TURN·uhl)—active at night

ocelli (oh·SEL·ee)—light-sensitive cells or eye spots found on all parts of the leech's body; groups of ocelli form eyes in the leech's head region

parasite (PAIR·uh·syt)—a plant or animal that lives in or on another living thing and gets food from it without giving anything in return; usually it harms and can eventually kill the host

predators (PREH·duh·turz)—animals that hunt and destroy other animals for food

proboscis (pruh·BAHS·uhs)—a sharp, hollow organ used like a straw that some types of leeches insert into their host to suck out blood

regenerate (rih·JEHN·uh·rayt)—to naturally replace a lost or injured organ or part by growing new cells

salivary glands (SAL·uh·vair·ee glandz)—glands that secrete saliva, a thin watery liquid that aids in digestion of food; in leeches, saliva often contains an anesthetic, an anticoagulant, and an antibiotic

sebum (SEE-buhm)—oily secretion of glands that open onto the skin

secretes (see-KREETZ)—forms and gives off

segmental receptors (sahg-MEHN-tuhl ree-SEHP-tohrs)—sensitive areas on the leech's body that detect light, water movement, vibrations, etc.; these sense organs help the leech find food

setae (SEE-tee)—small stiff bristles along each side of an earthworm's body which help it move and burrow into the ground; the Acanthobdellida, an unusual group of leeches, also have setae

spermatophore (spur-MAT-uh-fohr)—a small packet containing sperm; it is passed from one leech to another to fertilize the eggs of the other leech

suckers—the organs used by the leech for sucking and holding on to objects or surfaces

suturing (SOO-chur-ing)—stitching to close a wound

varicose veins (VAIR-uh-kohs VAYNZ)—abnormally swollen and twisted veins

BIBLIOGRAPHY

Books

Environmental Protection Agency. *Freshwater Leeches (Annelida: Hirudinea) of North America*, by Donald J. Klemm. Washington, D.C.: U.S. Government Printing Office, May 1972.

Johnson, Alvin J. *Johnson's New Natural History*. Vol. 2. New York: Henry G. Allen and Co., 1894.

Mann, Kenneth Henry. *Leeches (Hirudinea): Their Structure, Physiology, Ecology, and Embryology*. New York: Pergamon Press, 1962.

Muller, Kenneth J., John G. Nicholls, and Gunther S. Stent. *Neurobiology of the Leech*. Cold Spring Laboratory, 1981.

Nespojohn, Katherine V. *Worms*. New York: Franklin Watts, 1972.

Parker, T.J., and W.A. Haswell. *Textbook of Zoology*. London: McMillan and Co. Ltd., 1954.

Patent, Dorothy Hinshaw. *The World of Worms*. New York: Holiday House, 1978.

Sawyer, Roy T. *Leech Biology and Behavior.* Vol. 2. Oxford: Clarendon Press, 1986.

Tennent, Emerson. *The Natural History of Ceylon.* London: Longman, Green, Longman, and Roberts, 1861.

Journals

Altman, Lawrence K. "The Doctor's World." *The New York Times* (17 February 1981): C2.

"Bloodsuckers from France." *Time* (14 December 1981): 100

Branning, Timothy. "Learning from a Giant Leech." *National Wildlife* (April-May 1982): 35-37.

Brodfuehrer, Peter D., and Otto W. Friesen. "From Stimulation to Undulation: A Neural Pathway for the Control of Swimming in the Leech." *Science* 234 (21 November 1986): 1002-1004.

Budzynski, Andrei Z., et al. "Anticoagulant and Fibrinolytic Properties of Salivary Proteins from the Leech Haementeria ghilianii." *Proceedings of the Society for Experimental Biology and Medicine* 168 (1981): 266-275.

Budzynski, Andrei Z., Stephanie A. Olexa, and Roy T. Sawyer. "Composition of Salivary Gland Extracts from the Leech Haementeria ghilianii." *Proceedings of the Society for Experimental Biology and Medicine* 168 (1980); 259-265.

Clark, Matt, and Donna Foote (in London). "The Return of the Bloodsucker." *Newsweek* (2 February 1987).

Coniff, Richard. "The Little Suckers Have Made a Comeback." *Discover* (August 1987): 85-94.

"Diaries detail bizarre care Given Hitler." *Austin American-Statesman* (6 May 1983): A21.

"Doctors use hungry leeches to save boy's reattached ear." *Boston Globe* (25 September 1985).

Drotar, David Lee. "Who is Man's Best Friend? It Could Be the Lowly Leech." *Jack and Jill* (February 1982).

Foucher, G., et al. "Distal Digital Replantation: One of the Best Indications for Microsurgery." *International Journal of Microsurgery* 3 (4) (December 1981): 263-270.

Gasic, Gabriel J., et al. "Inhibition of Lung Tumor Coloni-
    zation by Leech Salivary Gland Extracts from Hae-
    menteria ghilianii." *Cancer Research* (April 1983):
    1633-1636.

Gibbs, Jerry. "Who Could Love a Leech?" *Outdoor Life*
    177 (April 1986): 34-35.

Gilette, Robert. "Leech Hunt: Success in the Swamp."
    Los Angeles Times (2 October 1977).

———. "Exotic Dangers Enliven Search for Giant Leech."
    Los Angeles Times (3 October 1977).

———. "In Search of a Giant Leech, French Guiana—A
    Step into Yesterday." *Los Angeles Times*
    (4 October 1977).

Gonzales, Arturo F., Jr. "Giving a Sucker an Even
    Break." *MD* (February 1987): 65-69.

Henderson, H. P., et al. "Avulsion of the Scalp treated by
    microvascular repair: The use of leeches for post-
    operative decongestion." *British Journal of Plastic
    Surgery* (1983).

Kennedy, J. Michael. "Leech Gains Hold as Aid to Surgeons." *Los Angeles Times* (16 June 1987).

"Leech Swimming: The neural story." *Science News* 130 (29 November 1986).

"Leeches Make a Comeback." *USA Today* 115 (October 1986): 14.

Moss, Michael. "U.S. Climbers Cross Nepal for try at Mount Everest." *Austin American-Statesman* (28 August 1983).

"Pharmacy Sells Leeches for 'Historical' Purposes." *Moneysworth* (February 1982).

Sawyer, Roy T. "Leeches, New Role for an Old Medicine." *Ward's Bulletin* (Spring 1986).

_____. "Why We Need to Save the Medicinal Leech." *Oryx* 16 (2) (October 1981).

Seigworth, Gilbert R., M.D. "Bloodletting Over the Centuries." *New York State Journal of Medicine* (December 1980): 2022-2027.

"The Bloodsucking Business." *Inc.* (May 1988): 14.

von Ranson, Jonathon. "Gay little creatures that nobody loves." *Yankee* (June 1981): 79-83.

Watchel, Paul Spencer. "Return of the Bloodsuckers." *International Wildlife* (September-October 1987).

Interviews

Baudet, Jacques, M.D. (Professor of Plastic and Reconstructive Surgery). Telephone interview with the author from the Hospital Du Tondo, Bordeaux, France, Summer 1983.

Gasic, Gabriel J., M.D. (Research professor, Laboratory of Experimental Oncology). Telephone interview with the author from the Pennsylvania Hospital, Philadelphia, Pennsylvania, August 1983.

Kristin, Bill, Ph.D. Telephone interview with the author from the Department of Zoology, University of California at San Diego, San Diego, California, 10 August 1983.

Malinconico, Scott, Ph.D. Telephone interview with the author from the Temple University School of Medicine, Philadelphia, Pennsylvania, July 1983.

Milas, Luke, M.D. (Head of the Department of Experimental Radiotherapy). Telephone interview with the author from the M.D. Anderson Hospital and Tumor Institute, Houston, Texas, September 1983.

Payton, Bryon, M.D. Ph.D. Interview with the author from the Department of Physiology, Memorial University of Newfoundland, St. Johns, Newfoundland, Canada, August 1983.

Sawyer, Roy T., Ph.D. (Managing Director, Biopharm, Ltd.). Telephone interview with the author from Swansea, Wales, June 1987.

Soileau, Carmen, M.A. (doctoral candidate, zoology). Telephone interview with the author from the University of Texas at Austin, Austin, Texas, Summer 1983.

Spicer, Thomas, M.D. Telephone interview with the author from the Department of Plastic and Reconstructive Surgery, University of Texas Health Science Center, Dallas, Texas, August 1983.

Strauch, M.D. (Professor and Chief, Division of Plastic and Reconstructive Surgery). Telephone interview with the author from the Albert Einstein College of Medicine and Montefiore Medical Center, Bronx, New York, 26 July 1983.

Walinsky, Michael (Pharmacist, Neff Surgical Pharmacy). Telephone interview with the author, Philadelphia, Pennsylvania, Summer 1983.

Weisblat, David, Ph.D. Telephone interview with the author from the Department of Zoology, The University of California at Berkeley, California, August 1983.

INDEX